HOW TO ANALYZE PEOPLE

SPEED READ PEOPLE AND ANALYZE BODY LANGUAGE & PERSONALITY

EDWARD BENEDICT

LEE DIGITAL LTD

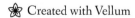 Created with Vellum

INTRODUCTION

Learning to analyze people, their behaviors, and their personalities can be very helpful to you, as it can strengthen the relationships you have with them. This relationship can be with anyone—whether your partner, your colleagues, your bosses, your family, or even your acquaintances. But why analyze people? Firstly, if we could only read people effectively, we would know if they liked something or if they are feeling comfortable or if they are agreeing to what we said. Secondly, it helps create empathy in you—and when you have empathy, you can handle crises and negative situations efficiently. It also helps you create better bonds with people.

In a professional capacity, knowing how to read people can get you far in the career ladder, especially if you have the capacity to understand the needs and wants of your client or boss. Working on your ability to read and analyze people can greatly affect how to deal with them, and this is especially important with people with whom you have a relationship—whether personally or professionally. When you understand and empathize on how someone else is feeling, you can adapt the way you convey your message and communication style so that the person you are communicating with can receive this message in the best possible way.

But how do you analyze people? What signs should you be looking into? What words must you listen to? What other signs can you target to help you understand what someone else is feeling and thinking? When reading or analyzing someone, one of the first few things you must realize is that you need to get rid of your biases and whatever apprehensions you may have made or have on them. These notions are merely walls that contain old and limiting ideas.

People who have mastered the art of analyzing other people are trained extremely well to analyze the invisible and use their senses to look deeper into a person's embodiment, looking further than where the general attention goes to. You've probably heard that the eyes are the window into the soul. If that's the case, then words are the gateway into the mind. When analyzing people, you not only need to look at the way they react, their body language, and even their eyes—but you also need to consider the kind of words they use to describe, explain, elaborate, and communicate. In this book, we will explore the proven strategies and steps on how to read and analyze people using non-verbal gestures and body language. The information in this book will help you how to study someone's body language, decipher the various personality types there are, and determine how you can communicate with them on a level that engages them. This book will also look into how to spot lies and deception as well as identifying romantic cues from a person.

CHAPTER 1

BODY LANGUAGE BASICS

BODY LANGUAGE IS one of the many ways that we humans communicate—but while most people pay attention to words and actions, body language is rarely given attention to. What we say sometimes can be very different from our body language. We may sound happy, but our body language says otherwise—and according to experts, our body language takes up half of the way we communicate.

Figuring out how to peruse non-verbal communication viably can enable us to comprehend what somebody is attempting to express, if they are comfortable with the choices they make and if they are genuine about them. We also learn to communicate our messages effectively when we learn to read body language. It goes beyond just what words can put meaning in to.

HOW COMMON IS BODY LANGUAGE IN COMMUNICATION?

Believe it or not but body language takes up 55% of our daily communication. However, analyzing nonverbal cues isn't focused on just the broad strokes. These gestures indicate various things, and it depends entirely on context.

Nonverbal cues are extremely crucial when trying to read someone because, in many ways, you can detect if someone is lying or if they are enjoying a date or how they are as a person when they come in for a job interview. It is tied with finding some hidden meaning to decipher non-verbal communication precisely so you know whether the individual's words are expressing how they really feel.

Unfortunately, we humans are more inclined to lie than to tell the truth for plenty of reasons such as avoiding conflict, trying to impress someone and so on. Sometimes, we end up lying more than once in a short span of time and while they may necessarily not be big lies, we end up doing it anyway. We end up willingly partaking in deception because we rather hear a sweet lie than the bitter truth. In any case, non-verbal communication isn't as misleading as words—our bodies are horrendous liars.

WHAT IS BODY LANGUAGE?

It is our body's physical, non-verbal communication approach that is sometimes in-sync with the words that are coming out of our mouth. Body language can be anything from a stance, an eye-glance, a quick facial expression and even the biting of our lip.

You may have seen how some people speak very animatedly and they use mostly their hands to convey or emphasize their words. There are a lot of hand talkers who keep their hands in a steady movement to pass on their point, stress on information, or just to keep the conversation going.

This gesture and many other forms of body language often speak volumes. Make an observation of how a person's body language is:

a) When they speak

- Do they have slumped shoulders? Is their back rounded with their head hanging down? This could indicate that they are either shy or sad.

b) When you see a person walking into a room to address a team or a company

- Do they carry themselves stably? Do they have their head at eye level or held high? This can be interpreted as arrogance or confidence.

c) When you need to talk to someone

- Do they have their arms folded across their chest? Do they have their legs crossed? Are they glancing around or sighing? This could be understood as an unfriendly stance, or they are not open to what you have to say, defensive or standoffish.

It is truly fascinating what body language can do and how much we can perceive from it. Aside from being able to utilize it in judging the mood a person is in or their attitude, you can also make and create better relationships simply by observing them. These non-verbal communication gateways help create a deeper sense of bonding.

BODY LANGUAGE BASICS

Your main goal when it comes to reading body language is to determine if a person is comfortable in the situation that they are currently in. Once you have established this, the next thing is to process the context that they are in and look at other cues. Of course, this is easier said than done so we will go into the specifics in following chapters. Here are some common denominators for positive body language:

- Extended periods of eye contact
- Feeling at ease
- Looking down and away out of shyness
- Moving or leaning closer to you
- Relaxed, uncrossed limbs
- Genuine smiles

Here are some common denominators for negative body language:

- Feet pointed away from you, or towards, and exit
- The feeling of unease
- Looking away to the side
- Moving or leaning away from you
- Rubbing/scratching their nose, eyes, or the back of their neck
- Crossed arms or legs

One body cue and mean plenty of different things. While crossed arms can be construed as negative non-verbal communication, it can likewise propose that an individual is feeling chilly, awkward, disappointed, or closed off. When understanding somebody, it is pivotal to focus on a few prompts since one's demeanor can be deceiving. It is a must to look deeper to understand what is really going on and this means focusing on cues and also the context it is in.

Here are some common body language categories:

BODY LANGUAGE CATEGORIES

Body language can be separated into general classifications:

1. Aggressive – Threatening body language
2. Attentiveness – This shows that you are interested and engages.
3. Bored – This is the complete reverse of attentiveness, and it is usually represented by not making enough eye contact and constant yawning.
4. Closed – This is when someone shuts you off and is often shown having crossed arms and standing far away.
5. Deceptiveness – This is usually portrayed when a person wants to get away with a lie and displays nervous behavior, guilty and worry.
6. Defensiveness – This person who is defensive can look like they are protecting or withholding information.
7. Dominant – Dominant body language is used when someone wants to be in command and they usually stand tall, with their chest puffed out.
8. Emotional – When a person is emotional, they are greatly influenced by their current emotions and usually have varying moods.
9. Evaluation – A person portrays a sense of evaluation in settling on a choice without second thoughts.

10. Greeting – It happens when two people first come into contact.
11. Open – This is, of course, welcoming and accepting.
12. Ready – It tells people that you are open, willing and prepared.
13. Content and relaxed – This can be portrayed by a calm, happy and restful demeanor.
14. Passionate – This is often a romantic body language that expresses attraction and is flirty.
15. Submissive – This shows off the relenting side.

The body language described above is usually commonly communicated through a combination of postures and poses and is not singular to one. Again, a wide range of body positions has altogether unique importance relying upon the specific situation, the circumstance, and the social foundation. Take for example the pose of crossing your arms—in a meeting situation, this can be construed as simply someone being serious and focused.

It is extremely crucial to take context into consideration and if you want to learn how to analyze people, then you need to have a heightened sense of awareness of how your body acts and what it is saying when you talk to the people around you. Keeping these tips in mind will help increase your communication and comprehension skills, thus opening a more effective line of communication with your team, your partner, children, and friends.

CHAPTER 2

UNDERSTANDING SELF

WHAT IS UNDERSTANDING SELF? This may or may not be the first time you have heard of this—but in the exploration and learning of analyzing people, the understanding self is one of the elements that we need to discover.

Understanding self is a spiritual and philosophical question which includes an in-depth study and journey. For example, when you are shown the forest or the deep blue sea, it is entirely up to you on how deep you would like to journey in and the kind of experience you want to derive from your journey.

UNDERSTANDING ONE'S SELF INVOLVES AWARENESS

For any ordinary human being, being self-aware all the time, 365 days a year, 24 hours a day is impossible and quite frankly, unrealistic. There is a certain level of awareness as well as a moment of realization. To be self-aware for even 10 minutes is already a great thing.

WHY IS SELF-AWARENESS HARD?

Don't worry if you do not feel like you understand or have the right or correct understanding on self-awareness. In fact, nobody ever really does. However, it is not too hard to try to understand once you learn it because self-awareness is about understanding your own self so you can be an effective and positive member of the community, your family, your company and generally at life.

The moment we are born into this world, we are already fed with information about the rights and wrongs of the world we live in by our parents, media, the education system and the society we live in. We also tend to consume plenty of information from the internet which makes not only focusing hard but also identifying who we are as a person.

Our body's main function is to survive which means that our body will do whatever it takes to survive, stay alive and be comfortable as much as possible. So when thinking about or understanding self, there are two things we need to look into which are:

1. Ego – which is the state of mind that is the most powerful. Egoist states are what rulers and conquerors have and will go to great lengths to ensure that their empire remains. Ego is the element to want everything which is why some people start spending or investing to accumulate wealth for generations to come.

2. Self – An ego mind does not let you become self-aware because self-awareness is spirituality and this type of spirituality has nothing whatsoever to do with religion. Spirituality is a way of life whereas religion is a lifestyle. The body wants its pleasure whereas the ego wants power, wealth and control.

A good leader of a country or state will lead a life that is not excessive and really, to live a life that is the exact same one as the average human in that country. What is the point of living in a large mansion or palace when the people in your country life in small squatters? If the purpose of a leader is to serve, then they should not have the entitlement of the state.

One must differentiate between ego and self because ego will blind a person from seeking righteousness and truth.

UNDERSTANDING YOURSELF

The list below can give you an idea of how to understand yourself but keep in mind that this list is as fluid and torrid as the ocean and you should look at yourself as the ocean because the value of Self is as deep as the big, blue ocean.

1. Desires and interests
2. Your needs and wants
3. Passion
4. Identity – country, race, language, clan, community, sexual preference
5. Emotions
6. Awareness to thoughts
7. Ideas
8. Relationships
9. Your strengths and weakness
10. Skill
11. Knowledge
12. Expertise

Understanding self is a process or exercise that you need to do every day. It is about peeling off your superficial layers and getting lighter as you peel each layer away and feel the peace emanating from within. Understanding self also does not mean that you are forgoing your desires or ambitions whether personally or professionally.

Your motivation increases when you take on a task or project when you begin to understand yourself. However, do take note that it is not as easy as A, B & C.

As you go on the journey of self-discovery and seeking yourself, the truth about yourself will surface bit by bit. Think of it as mining diamonds. If diamonds can be found anywhere, they lose their value but because they are rare and exquisite, it is considered precious and demands a high premium. Self-Awareness and understanding self is exactly like the process of mining diamonds —it comes to you a little by little as you journey through life and discover a piece of you. You will value this piece more profoundly.

UNDERSTANDING SELF FOR YOUR CAREER DEVELOPMENT

Knowing yourself not only extends value to your relationships but it also enhances the partnerships and your ethics in your career. By being self-aware, it can help you make thoughtful career decisions and plan a career development path suitable for you. Pivotal as well for you is establishing your career goals and things that provide you with joy, fulfillment, as well as passion. Self-awareness also means pinpointing what kind of environments and work cultures that you thrive in because all of these elements will enable you to focus on a career path that has many opportunities that have the kind of environment and culture that will help you thrive, grow, and learn.

Some of us prefer routine with very minimal disruption. We like order, routine, going in at a certain time and leaving at a certain time because it enables us to pursue other passions and obligations. Some of us prefer an environment that is fast paced, full of energy and unpredictability.

Uncovering your Self

REALIZING YOUR TALENTS

To unlock what your talents are—if you don't know it already or having fully explored it—you can:

- Think about the significant achievements in your life and why you achieved it.
- Challenge what you have in life now.
- Question why these elements are significant.
- Rundown down what you have gained from them and how they impact you and the profession trajectory that you seek.
- Understand your personality classification.

Of course, someone's personality gets the biggest chunk in understanding self simply because your personality can provide the elements needed to establishing and perceiving why you have inclinations to act or respond with a specific goal in mind and help distinguish the sorts of workplaces you flourish in. To find out the classification that best describes you, it is always possible to look for tests such as the Myers Briggs Type Indicator. Doing this can help you understand yourself better and work towards refining your career goals so it is more aligned to what you like, what you can do, what you are interested in and where your passions life

UNDERSTANDING YOUR VALUES AND MOTIVATIONS

These are also indicators of a successful career development. We all get motivated based on various different things although money is the most common of all. But what other motivations would you consider as part of the factors for you strive to do better? What are these elements that you could consider form the foundation of your career success? Understanding these elements would help you find and stay in a job that you feel satisfied at the end of the day.

AUDITING YOUR CAPABILITIES AND EXPERTISE

Part of understanding one's self is also understanding and exploring what you are good at, where your strengths lie and what are your fortes. What aptitudes and characteristics do you have because of your examination, side interests, or even paid or intentional work involvement? If you want to find out, you can utilize the Vitae Researcher Development Framework (RDF) to map your current competencies, attributes, and capabilities.

UNDERSTANDING YOUR LEARNING STYLE

The path towards career development is paved with plenty of learning opportunities so identifying your most effective and procrastination-free learning capacity can help you pinpoint what kind of training or course you can and should undertake so that you can develop your capabilities and expertise efficiently and effectively. Again, you can use a variety of tools in determining the learning style that best suits you—and a powerful one is the Learning Style Questionnaire (Honey and Mumford, 1982).

THE IMPORTANCE OF UNDERSTANDING SELF

So what happens when you know yourself? Why is it so important to know yourself?

Here are some of the benefits:

- You are happier -It's a fact. You will be happier when you can express who you are in a song, in art, in music, in words. When you express your desires you will also be more likely to get what you want.
- You have fewer conflicts – These conflicts are more internal conflicts that you have with your inner self. When your actions outside are in tandem with your internal feelings and beliefs and values, you have lesser inner conflicts to deal with.
- You are better at decision making – Knowing self means you are also sure of yourself and this makes it easier for you to make better, informed choices about everything that happens in your own life. This can be anything from small decisions like what color to paint your walls too big decisions like who to marry. Your internal compass gives you a set of invisible guidelines that your mind processes to help you solve these issues.
- You gain better self-control – You have a better ability to resist bad habits simply because you know yourself better. You are also inclined to develop good ones. Self-control also helps you gain better insight to know which of your values and goals that activates your willpower.
- You rarely give into societal pressure – Knowing yourself also means your values are grounded and you

know what your preferences are. You will be firm with
your NO rather than let peer or social pressure give
into you saying yes.

- Better tolerance and understanding – Your awareness
  of your own struggles and foibles make you more
  perceptive of other people's needs. You can also
  emphasize better with the problems and needs of other
  people.
- Vitality and pleasure – Being you in all its simplicity
  and truthfulness enable you to be more alive and also
  enables you to engage and experience life on a larger,
  richer and more exciting scale.

STOIC VALUES AND IDEAS IN UNDERSTANDING SELF

Value

The only one good thing is an excellent mental state which is
rationalized by ethics and reason, and this is the main thing that
can ensure an individual's everlasting bliss. Money, success, fame,
and other material items may only give us temporary joy but make
it hollow inside. There is nothing wrong with any of these – you
need things like this to survive in this world and form a good life
but excessive pursuit of these things will only lead to long-term
damages. We will not feel a sense of fulfillment. Fulfillment and
happiness can only be achieved through a rational state of mental
excellence.

Emotions

Our judgments are the architects of our emotions. This judgment, is how we think something is good or bad that is going to
happen or already happening. Oftentimes, our mistaken judgments are the cause of negative emotions. However, as we know in
Stoicism, the only things we can control our own thoughts and

actions. So keeping this in mind, these judgments are within our control which means so is our negative emotions. Stoicism, despite the popular opinion that it is an emotionless school of philosophy, is anything but emotionless. Stoicism principles are not repressive or denying anyone of their emotions. Stoicism instead views emotions are something entirely different. In following Stoic principles, you do need overcome negative and harmful emotions that are a result of mistaken judgment and at the same time, embracing the positive emotions. For example, you replace anger with happiness.

Nature

Living in agreement or in harmony with Nature is the Stoic principle of attaining the Good Life. We need to acknowledge that we are all but a small entity of the large universe. We are all small cogs in the big scheme of life and in order to live the Good Life is to live according to or in line with nature.

Nothing can be picked up from endeavoring to oppose these bigger procedures with the exception of displeasure, dissatisfaction, and frustration. While there are numerous things on the planet that we can change, there are numerous others we can't—and we have to comprehend this and acknowledge it.

Control

One primary principle that Stoicism rallies are the things in the life of what we can control and what we cannot control. Our thoughts, feelings, emotions, desires, and actions are all things that are within our control. The external processes, other people's reactions, the results, consequences, objects are all outside our control. Our unhappiness, most of it at least is a result of confusing these two categories – Thinking of the things that we have control but in reality, we do not. Thankfully, one thing that we happily do have control is the guarantee of a good, happy life.

The three popular Roman Stoics—Marcus Aurelius, Seneca

and Epictetus offers a wide range of advice that is practical even for modern society. These writings were written with the sole purpose of helping people in their everyday lives and understanding self-worth. Seneca in the later years of his life, living quietly in his country estates, wrote two of his most famous books, the *Naturales Quaestiones*—an encyclopedia of the natural world; and his Letters to Lucilius, which details his philosophical ideas and thoughts.

Marcus Aurelius was also known as the philosopher king, until his death. He wrote the book Meditations as his own journal for self-improvement and guidance. Epictetus, born a slave but eventually rose to become a great Greek Stoic philosopher. His teachings and philosophy were written down and published by his pupil in books known as Discourses and Enchiridion.

CHAPTER 3

CLUES TO REVEALING TRUE INTENTIONS - EYES

THE EYES ARE DEFINITELY the windows to our soul—don't you agree? When we see a person for the first time, our gaze automatically goes to their eyes—looking, searching, and wondering who this person is.

In all honesty, it is easier to evaluate a person's heart than their mind. We can effortlessly pick up on our friend's mood or sense why our partner has dismissed plans to meet even without them speaking a word. How do we know this? How do we know what is going on in their heads without even speaking a word to them? For close friends, our partners, brothers, sisters, and family members—we just know simply because we grew up with them or because we've known them for a considerable amount of time to know what floats their boat, so to speak.

But how do we get this special access to the human mind towards acquaintances or your colleagues? Recent research tells us that looking at people's eyes is one of the ways to get in touch with the human mind, hence the phrase "I can see it in your eyes." It is definitely poetic, and that's why you see it in so many music lyrics.

While it's all beautiful and romantic, the truth is that the eyes can tell a lot about a person because while people can somehow hide their emotions and check their body language, they can't change the way their eyes behave.

HOW DOES THE LANGUAGE OF THE EYES WORK?

When studying a person by looking in their eyes, firstly you need to do is subtly and not stare into their eyes. You need to maintain eye contact in a friendly manner and when you have established this, look into the changes in the pupil size.

A popular study published in 1960 says that the wideness or narrowness of pupils reflects how certain information is processed and how the viewer finds it relevant. The experiment was conducted by psychologists Polt and Hess from the University of Chicago, who analyzed both female and male participants when they looked at semi-nude images of both sexes. The study showed that female participants pupil sizes increased in response when they viewed images of men and for the male participants, the pupil sizes increased when they viewed images of women.

Hess and Polt in subsequent studies also found that homo-sexual participants looking at semi-nude images of men (but not of women) also had larger pupils. This is no surprise at all because pupils also reflect how aroused we are. Women's pupils responded to images of mothers holding babies. This goes to show that pupil sizes do not reflect how aroused we are but also how we find a piece of information relevant and interesting.

This idea was brought forward by Daniel Kahneman who led a study in 1966. Kahneman is now a Nobel-prize winning psychologist. His study required participants to remember several three to seven digit numbers and participants were to recall it back after two seconds. The longer the string of digits was, the larger their pupil sizes increased which also suggested that pupil size was also related to the information that the brain processing.

In looking for clues in the eyes of a person, the first step is to know what that person is thinking and to look deeply in their eyes.

Apart from the processing of crude information, our eyes can also send more sensitive signals that other people can pick up, especially if they are extremely intuitive. Another study conducted by David Lee began by showing participants images of other people's eyes and he asked them to determine what kind of emotions this person was experiencing. This researcher from the University of Colorado found that participants could correctly gauge the emotions whether it was anger or fear or sadness just by looking at the eyes.

The eyes also have the ability to reveal much more complex phenomena such as whether a person is telling the truth or if they are lying. For example, Andrea Webb conducted a study in 2009 which had one group of participants steal $20 from a secretary's purse and another control group was asked not to steal anything. This research led by the Webb and her colleagues from the University of Utah showed that pupil dilation gave away the thief. All participants were asked to deny the theft and the analysis of pupil dilation showed that participants who lied had pupils that were one larger by one millimeter compared to the pupils of participants who did not steal.

Our eyes also have the capacity to become a good indicator of what people like. To learn to read the signs, you would need to look at the size of the pupil as well the direction of gaze. Take for example someone choosing what they would like to eat at a restaurant. We are visual creatures anyway so our eyes are most likely darting between choosing the salad or the cheeseburger.

The other point to look into is decision making. When we are making a difficult decision, our eyes tend to switch back and forth between the different options in front of us and our gaze ends at the option that we have chosen. By observing these little details of where someone is looking, we can identify which options they choose.

Another way of studying this type of difficult trade-off is offering monetary bets to participants. A study conducted in Brown University by James Cavanagh was when participants were asked questions that involved difficult tradeoffs between probabilities and payoffs.

Participants were paid based on their decisions. The researchers were kind that the harder the decisions were, the more the pupils of the participants dilated. As the choices got harder, our pupils also got bigger.

The eyes also give away clues to if we experienced something unpleasant. Another study on eyes and its reaction was conducted in the University of Washington in 1999. The painful simulation was administered to the fingers of 20 participants and they were asked to rate this pain from tolerable to intolerable. The more intolerable the circumstances were, the larger the pupils of the participants became.

Although pain is a very different feeling than looking at images of seminude people, it still showed a change in pupil response. This shows that pupil size correlated with the strength of feelings and whether those feelings were positive or negative. So if you want to know whether a person is feeling bad or good, consider the context and look into their eyes.

SO WHAT DOES THIS MEAN IN TERMS OF NON-VERBAL COMMUNICATION?

Can we read everything just by looking at the eyes? Are the eyes the only signals we should concentrate on?

The thing is, the eyes are just one of the indicators or signals that we communicate with. When making high-stakes decision duh as whether a person is guilty of a crime, pupil dilation is not something you solely rely on to make a judgment.

We should also look into context. That said, we are more perceptive to the body language of the people we always come into contact with compared to total strangers simply because we can tell their regular facial expressions apart from the non-regular ones.

To make better assessments if feelings, we need to look at various other evidence or elements of body language and of course context. Because people cannot change how their pupils behave, the eyes are often used as a source of information to help create a better relationship simply because it enables us to empathize better. You may not be able to read a person's exact thoughts just by looking at their eyes but it still is a good perspective to study body language and read people.

CHAPTER 4

CONTEXT – THE CUES THAT TELL IT ALL

IN THE PREVIOUS CHAPTER, we talked about how when looking at body language, taking note of the context was also crucial. For example, say, you are having dinner with a friend, and they start fidgeting or have their arms crossed, their body slumped, and their look seemingly bored. There are several possible explanations for this.

For one, they could be uncomfortable with the conversation topic. Alternatively, they could be uncomfortable because you're a loud eater, and they are not sure if they should tell you—or it just could be they are cold and feel uncomfortable sitting in the chair for too long. If you only looked at their body language and deduced that they are uncomfortable because of you, or what you are talking about, you will not get the right answer as to why they are uncomfortable. Body language will only tell you if someone *is* comfortable or uncomfortable—but it will not tell you *why*.

This is why you should look at *context*.

WHAT IS CONTEXT?

Context is the surrounding events and occurrences. It simply means taking into consideration the circumstances forming the current situation, the background events, idea or statement that is taking place or took place moments ago. Context enables the viewer or reader to comprehend the situation.

When looking at the context in body language, you need to be aware of three things:

1 – The conversation that is taking place – did something said in the conversation cause the person to become even more or less comfortable? Did their language change when you asked them a specific question or mentioned a specific statement? It could be something in the conversation that made them uncomfortable.

2 – The environment where the conversation takes place – Look around the area you are in when having this conversation. Is it in public? Is it in an open space? Are people around? Or is it in a private place where nobody can see the both of you? Both open or closed spaces can make a person uncomfortable depending on what is being discussed.

3 – The recent experiences of the person you are speaking to – This person's day may or may not have started with you talking to them. The previous day's experience may still be affecting this person. It could be that they had a rough day at work or they are not well or they are just stressed out.

APPLYING CONTEXT

When analyzing people, you need to take into account the context that they are in. Taking time to look at this element will enable you to identify potential causes of their discomfort. When you remove this discomfort, such as moving to a different room or speaking in more hushed tones changing the subject, you can see the difference it makes towards this person.

Take for example if their body language showed you some sense of discomfort when you start talking about a controversial topic. Change this topic and see if they look a little relaxed. If you still cannot pinpoint the source of their discomfort, just remember that the best way to find out is just to ask them.

Even if you do not know the source of discomfort, you can still attempt to make changes so that they are more comfortable. You can change the smallest things to create a different spark. For example, offering the person a drink before you give them some bad news or even asking about their day to make them feel more relaxed before starting an interview. While it is important to know the source of discomfort, simply just wanting to find out or even being away of their discomfort and trying to make it better would make the whole situation go a long way from bad to good.

IDENTIFYING CONTEXT

It will take some time to practice finding out context before you are comfortable with looking for it and also focusing on the conversation that is happening. But if you always think about context when a conversation is happening, you are one step closer to practicing context.

PURPOSE OF LOOKING FOR CONTEXT

The purpose of identifying context is to help you look for clues that will make the person you are speaking to more comfortable. When a person's body language shows you that they are uncomfortable, it helps that you look into context surrounding the conversation. This not only builds better rapport but it also helps you empathize better. Practice looking at context by consciously looking at your surroundings, the topic of the conversation and even your tone of voice.

Remember that even your body language plays a big part in identifying context, from the way you move, the tone of voice you use and even eye contact. It isn't just about the person's body language.

CHAPTER 5

BEHAVIOR ANALYSIS

WHEN TALKING ABOUT BEHAVIOR ANALYSIS, we focus on using learning principles to bring behavioral change. This is actually a branch of psychology that aims to understand the unforeseen cognitions and focuses on the behavior of a person and not on the mental causes of said behavior.

Behavior analysis has extremely fruitful practical applications when it comes to mental clarity and health, especially in helping children and adults learn a new sense of behaviors or reduce certain problematic behaviors.

ANALYZING SOCIAL BEHAVIORS

The intentions behind certain actions of ours are commonly hidden. When a person is feeling angry or feeling depressed for example, their behavior portraying this feeling is usually very different such as they would keep quiet or go for a smoke to calm down.

Another example is also the kind of words used to convey

dissatisfaction such as 'sure, go ahead' or 'fine' when actually we are not fine with the solution or the decision made. Empathy is a much-needed when it comes to analyzing social behaviors such as this because, at the end of the day, you want to understand and listen and not just hear to answer.

In analyzing behaviors, demonstrating trust and building rapport is extremely crucial because when you display empathy, you naturally break down any subversions and focus on the heart of the matter.

A rule to remember is that when you do experience emotions and feelings, you must know that people around you will not know about it unless they sense a change in your body language. When nobody understands or get it, there is no need to get angry—but of course, it is easier said than done. We get angry when nobody, especially someone close to us doesn't notice that we are angry.

BEHAVIOR IS LARGELY DICTATED BY SELFISH ALTRUISM.

Nobody is completely selfish and if we were to make such a claim, that would mean we are totally ignoring the acts of sacrifice, kindness, and love that goes around the world. However, most behavior does come out from the elements of selfish altruism.

It is a win/win situation when it comes to selfish altruism. It is a basic two-way road of you help me, I will help you. Here are a few scenarios where selfish altruism applies:

1. Transactions: If you were to purchase a car, both the seller and the buyer mutually benefit. The buyer gets the vehicle; the seller gets their sales. This is a primary form of selfish altruism between two people who do not have any kind of emotional bonds.
2. Familial: Our mind is designed to protect the people

with whom we share our genes with. We have a higher tendency to protect these people and this sense of protectiveness depends on close friends to loved ones to siblings and family.

3. Status: People sometimes, not all the time, help someone as a sign of power. Sometimes, people offer assistance and help to boost their reputation and self-esteem.

4. Implied Reciprocity: Plenty of relationships are based on the fact that if I offer you with assistance one day, you would remember it and help me out as well one day when I need it.

Some certain behaviors are not part of the categories described above. For example, nameless heroes dying for a cause that does not directly benefit their country or bloodline. Another example is volunteers who devote their time selflessly towards missions and aids. But of course, these are just the smaller portion of the entire world community. The motives of people and what appeals to them is what you need to understand. When you do, you find ways to help people within these four categories. It's very rare not to expect people to give aid that does not benefit them in any form or way.

PEOPLE HAVE POOR MEMORIES.

Not everyone has a bad memory, but our minds have both long-term memory storage and short-term memory storage. For example, ever been introduced to someone at a party and then you just forgot their name the day after? People have trouble remembering things, especially something not relevant to be stored in their long-term memory. People are more likely to remember similarities that they share with you rather than differences.

When analyzing people, remember that people generally forget things so do not assume that they are disinterested with the information you have given or have malice against you.

PEOPLE ARE EMOTIONAL.

People have stronger feelings about certain things more than they let on but they can't show their specific emotions too much, especially negative emotions such as anger, outbursts, and depression simply because it is generally frowned upon by society. The rule is not to assume everything is fine just because someone isn't having an outburst. Sometimes the strongest ones are the ones that suffer most. All of us have some form of a problem, and these issues are normally contained. You necessarily do not need to call people out on their private deception, but what you need to do is be a little sensitive to those unseen currents and empathize with people because this gives you an advantage when you are trying to help.

PEOPLE ARE LONELY.

When you look at all these Instagram influencers having the time of their life or even celebrities going for numerous parties and ceremonies, the last thing you'd think is that they are lonely. The reality is, many people who seem like they have it all are actually quite lonely. People are sensitive to threats of being left out or ostracized or even having the fear of missing out. Loneliness and the desire to be among people exist in all of us, even if we are introvert. Analyzing this behavior is knowing that loneliness is very common among people and in this sense, you're not alone in feeling this way.

PEOPLE ARE SELF-ABSORBED.

Like it or not, people tend to be more concerned about themselves than about other people. Just look at social media and you can see how self-absorbed people are especially with an account full of selfies. People are more concerned more about themselves to give you any attention and for people to be lonelier, more emotional and feel different than they let on depends also on how you see the world. This perspective makes you independent and also proactive at the same time when you think about it. You become independent so you do not have to rely on anyone and you are more proactive so you have things to do and places to go on your own without depending on enjoying good times with other people. You place your own individual happiness in your own hand rather than in the hands of other people.

When analyzing people, just remember that in some ways or another, they all think and act like you in varying degrees.

CHAPTER 6

WHAT IS VERBAL COMMUNICATION?

VERBAL COMMUNICATION IS a form of communication that uses languages and sounds to convey a message. Verbal communication serves as a channel to express your ideas, desires, and concepts—and it is extremely crucial to the process of teaching and learning. Verbal communication is used in tandem with nonverbal forms of communication—however, as humans, we use verbal more than any form of communication when talking to people.

DEFINING VERBAL COMMUNICATION

When people look at the written word, they often relate it to the action of talking. No matter what civilization we come from, humans have always used verbal communication as a means to exchange thoughts and ideas and messages. Verbal communication is always spoken communication but written communication is also a form of verbal communication. For example, when you read this book, you are decoding the writer's point of view to learn more. In this chapter, we will explore the various elements that define verbal communication and how it affects our lives.

THE BASICS

As we know, verbal communication is all about the written and spoken word. It refers to the words we use when we speak or write whereas nonverbal communication is all about communication that takes place using other forms of communication other than words and body language is one such way that we humans use to communicate through gestures and well, silence too.

Here is a great table that helps simplify the understanding of both verbal and nonverbal communication.

VERBAL COMMUNICATION
Nonverbal Communication
Oral
Spoken Language
Laughing, Crying, Coughing, etc.
Non-Oral
Written Language/Sign Language
Gestures, Body Language, etc.

TYPES OF VERBAL COMMUNICATION

Public speaking and interpersonal communication are the most common types of verbal communication. Public speaking generally refers to any kind of communication done verbally to a group of people whereas interpersonal communication refers to an exchange that takes place involving a group of people that are simultaneously talking and listening.

PROFESSOR ROBERT M. KRAUSS FROM COLUMBIA UNIVERSITY says that signs and images are the essential flags that establish verbal correspondence. Words go about as images, though signs are optional items that help the hidden message and incorporate things with the manner of speaking and even outward appearances.

THE PURPOSE OF VERBAL COMMUNICATION

It is important no doubt, as without verbal communication, you wouldn't even be reading this book. But apart from that, verbal communication has plenty of purposes chief of it being to relay and convey messages to at least one of the receivers. It incorporates everything from one-syllable words and sounds too complex sentences and dialogs that depend on both feeling and dialect to create the ideal result and impact. Verbal correspondence is utilized to ask, advise, contend, examine, present and spread subjects of different sorts. In teaching and learning, verbal communication is extremely important. It also bonds and builds relationships and of course used wrongly, destroys and breaks relationships too. While every human on earth communicates in one form or another, the language we speak is a human phenomenon that enables us to convey our messages in a more precise way that any other forms of communication used by other beings such as animals.

THE CHALLENGES OF VERBAL COMMUNICATION

Challenges do happen as we all know with verbal communication especially when we are trying to express ourselves. Misunderstandings happen because we use poor word choices, arguments may arise when two or three people have clashing perspectives and when faulty communication techniques are used, sometimes the messages conveyed is understood wrongly and all of these causes a breakdown in communication. Also, knowing and speaking different languages also become a barrier to communication can cause confusion, which is why nonverbal communication is used.

PREVENTING COMMUNICATION BREAKDOWN

While confusion and misunderstanding do occur and it cannot be avoided completely, we can, however, choose to communicate effectively to lessen these disrupt. Always think before you speak is one golden rule, also consider the message you want to convey before speaking about it and also take into consideration the recipient's point of view. Not everyone is going to agree with you. Also paying attention to nonverbal cues and body language will help you craft what to say and how to say it—depending on the situation at hand. Always enunciate and speak your words clearly when you communicate.

While everyone has a different style of communicating and perceiving messages, you can only control what comes out of your mouth so only you can craft and convey your message as well as possible to ensure those listening to you understand what you are trying to say. Using nonverbal actions in your communication can also greatly affect the way the message is understood and perceived.

A SYSTEM OF SYMBOLS

Symbols are also worth exploring in this chapter of verbal communication. What is symbolled? Nelson & Kessler Shaw describes symbols as ideas, thoughts, emotions, objects and even actions that are used to decipher and give meaning. Symbols represent or stand for something. It serves as a symbolic representation of the idea of a sentence we want to say and also sometimes for an object.

CHARACTERISTICS OF SYMBOLS

- Symbols are Arbitrary – The symbols used are arbitrary, and they sometimes have no direct relations to the ideas that represent or the objects we want to describe. However, using symbols is considered successful communication especially when we reach an agreement on the meaning of these symbols.

- Symbols are ambiguous – Symbols have several possible meanings. For example, a Blackberry in today's world can be anything. It can be the fruit or it can be a mobile phone. Or take Apple for instance – it can be a fruit or a computer or if someone says they were on a date with someone really cool, does that mean that person is an awesome person or they are cold? Or if someone says they are gay, does that mean they are happy or they are homosexual? Meanings towards symbols change over time because of the changes happening in the world today, the shifting social norms, the advances in technology. We are all able to communicate in symbols because of the finite list of possible meanings and these meanings relate to the members of a given language agree upon. Without this agreed-upon symbolism, we share very little meaning and context with one another.

- Symbols are also abstract – They are not material or physical. A specific dimension of deliberation is a fundamental component in the way that images just establish of thoughts and items. This abstraction enables us to use phrases such as 'the public' to mean the people from a certain state or place rather than having to be distinctive about the people in a country based on their diversity. For example, the non-wizarding world in Harry Potter is called 'muggles' rather than explaining the separate culture of muggles. Abstraction is useful when you want to convey an intricate concept in a simple, straightforward way.

CHAPTER 7

COMMON PATTERNS OF INTERPRETING BEHAVIOR

HUMAN BEHAVIOR IS A COMPLEX THING. Because of its complexity, reading and analyzing people is not as easy as it sounds—but neither is it hard simply because as human beings, we exhibit more or less the same kinds of mannerisms and behavior when we experience a certain emotion or action.

SO WHAT EXACTLY IS BEHAVIOR?

Essentially, scientists categorize human behavior into three components:

- actions
- cognition
- emotions

ACTIONS ARE BEHAVIOR.

An action is regarded as everything that constitutes movement and observation whether using your eyes or using physiological sensors. Think of actions as a form of transition or even an initiation from one situation to another. When it comes to behavioral actions, these can take place at different scales and they range from sweat gland activity, sleep or food consumption.

COGNITIONS ARE BEHAVIOR.

Cognitions are described as mental images that are imprinted in our minds and these images are both nonverbal and verbal. Verbal cognitions are such as thinking 'Wow, I wonder what it's like to wear a $2000-dollar designer dress' or 'I have to get the groceries done later' all constitute verbal cognition. However, imagining things, in contrast, is considered nonverbal cognition, such as how your body will look after losing weight or how your house will be after a repaint. Cognition is a combination of knowledge and skills and knowing how to skillfully use them without hurting yourself.

EMOTIONS ARE BEHAVIOR.

An emotion is basically any brief conscious experience that is categorized by an intense mental activity and this feeling is not categorized as a coming from either knowledge or reasoning. This emotion commonly occurs or exists on a scale starting with positive vibes such as pleasurable to negative vibes such as being unpleasant. There are other elements of physiology that indicate emotional processing—such as an increase in respiration rate, retina dilation, and even increase in heart rat—all a result of increased or heightened arousal. These elements are usually invisible to the naked eye. Emotions, similar to cognitions also cannot be noticeable to the naked eye. These can only be noticed through tracking facial electromyographic activity (fEMG) indirectly which monitors the arousal using ECG, analyzes facial expressions, respiration sensors, galvanic skin response as well as other self-reported measures.

EVERYTHING IS CONNECTED

Cognitions, emotions, and actions run together and simultaneously with one another. This excellent synergy enables us to understand the events, activities, and happenings that are happening around us, to get in touch with our internal beliefs and desires and to correctly or appropriately respond to people that are in this scenario.

IT IS NOT THAT EASY TO UNDERSTAND AND DETERMINE WHAT exactly is the effect and cause. For example, when you turn your head, which is an action and seeing a face familiar to you, this will cause a burst of joy, which is the emotion and is usually accompanied by the realization which is the cognition. In other words, it is through this equation:

ACTION = EMOTION (JOY) + COGNITION (REALIZATION)

IN SOME OTHER SCENARIOS, THIS CHAIN OF EFFECT AND cause can also be reversed – you may be sad (experiencing an emotion) and you proceed to contemplate on relationship concerns (you go through cognition) and then you proceed to go for a run to clear your mind (you take an action). In this case, the equation would be:

EMOTION (SADNESS) **+** *COGNITION (**I** NEED TO GO FOR A run) = action*

CONSCIOUS **+** UNCONSCIOUS BEHAVIOR

CONSCIOUSNESS IS AN AWARENESS OF OUR INTERNAL thoughts and feelings and it also has to do with proper perception for and the processing of information gathered from our surroundings. A big portion of our behaviors is through the guided unconscious processes that surround us. Like an iceberg, there is a huge amount of hidden information and only a small fraction of it is obvious to our naked eye.

OVERT **+** COVERT BEHAVIOR

OVERT BEHAVIOR FOCUSES ON THE ASPECTS OF BEHAVIOR which can be observed by the naked eye. These behaviors are such as body movements, or as some would call it–interactions. Physiological processes such as facial expressions, blushing, smiling and pupil dilation may be subtle but it all can still be seen. Covert expressions are thoughts or cognition, feelings which are emotions and responses that are not easily or visibly seen. These subtle changes in our body's responses are usually not seen by the observer's eye.

IF WE WANT TO OBSERVE COVERT RESPONSES, THEN physiological or biometric sensors are usually used to help in observing them. As mentioned earlier on in this chapter, usually EEGs, MEG, fMRI or fNIRS are all used to look out for physiological processes that showcase covert mannerisms and behavior.

Rational + irrational behaviors

ANY ACTION, COGNITION OR EMOTION WHICH IS GUIDED OR influenced by reason is considered rational behavior. Irrational behavior, in contrast, is any action, emotion or cognition that is not objectively logical. For example, people who have extreme phobias are considered as having irrational fears, which are fears that are cause them to behave a certain way.

Voluntary + involuntary behaviors

WHEN AN ACTION IS SELF-DETERMINED OR DRIVEN BY decisions and desires, this is often categorized as voluntary actions. Involuntary on the other action would be actions that are done without intent, by force or done in an attempt to prevent it. People who are in cognitive-behavioral psychotherapy are often exposed to problematic scenarios involuntary as a form of therapy so that they can help get through this fear with the help of the therapist at hand. Now that we have a form of understanding of human behaviors, here is how we can interpret these behaviors. Keep in mind that these are just the surface or basic ways that interpretation can be done as there are more other complex and detailed ways.

#1 Establish a baseline

When you read people, you would notice that they all have unique patterns and quirks of behavior. Some people look at the floor while talking, or they have a habit of crossing their arms, some clear their throat ever so often while some pout, jiggle, or squint even. These behaviors are displayed for various different reasons – as we concluded in chapter 1, it could be they are uncomfortable or it just could be a habit. However, these actions could also mean anger, deception or nervousness. When reading people, we first need to form a baseline by understanding context and also what is normal behavior for this person.

#2 Look for behavior deviations

When you have established baseline behaviors, your next goal is to pay close attention to the inconsistencies that show up between the baseline mannerisms and the person's words and gestures. Say for example you've noticed that your teammate usually twirls their hair when they are nervous. As your teammate starts their presentation, they start to do this. Is this common behavior in your teammate's mannerisms or is there more than meets the eye? You might want to do a little bit more digging and probe a little bit more than you normally would.

#3 Start noticing a collection of gestures

A solitary word or gesture does not necessarily mean anything but when there are a few behavioral patterns start forming, you need to pay attention to them. It could be that your teammate starts clearing their throat in combination to twirling their hair. Or they keep shifting. This is where you need to proceed with caution.

#4 Compare and contrast

So we go back to the teammate again and you've noticed that they are acting more odd than usual. You move your observation a little closer to see when and if your teammate repeats this behavior with other people in your group. Make an observation on how they interact with the rest of the people in the room and how their expression changes, if at all. Look at their body language and their posture.

#5 Reflect

This reflection isn't about meditation rather it is to reflect the other person's state of mind. As human beings, we have mirror neurons that act like built-in monitors wired to read another person's body language simply because we also have these mannerisms as well. For example, a smile activates the smile muscles in our faces whereas a frown activates the frown muscles. When we see someone that we like, our facial muscles relax, our eyebrows arch, our blood flows to our lips making them full and our head tilts. However, if your partner does not mirror these set of behavior, then it could be that they are sending a clear message which is they are not as happy to see you.

#6 Identifying the resonant voice

You might think that the most powerful person is the one that sits at the head of the table or the one that is standing in the front. That is not always the case. The most confident person always has a stronger voice and they are more likely the most powerful one. Just by looking at them, you can deduce they have an expansive posture, they have a big smile and a strong voice. However, make no mistake that a loud voice is not a strong one. If you are presenting to an audience or pitching an idea to a group of people, you would normally focus on the leader. What happens when the leader has a weak personality? They will depend on others to make a decision and they are easily influenced by them. So when pitching or presenting to a group, identify the strong voice and you'll have a stronger chance of success.

#7 Observe how they walk

People who shuffle along or lack a flowing motion in their movements or always keep their head down while walking lack self-confidence. If you see this exhibited by a member of your team, you might be inclined to make extra effort to recognize their contribution in order to build this person's confidence. You might also need to ask them more direct questions at meeting so that they are inclined to offer their ideas out in the open as opposed to keeping them quiet.

#8 Using action words

Words are usually the closest way for people to understand what is going on in another person's mind. These words symbolize the thoughts that are running through their mind and in identifying these word, you also identify its meaning. Say for example if your friend says 'I decided to make this work', the action word used here is 'Decided'. This solitary word shows that your friend is 1 – not impulsive, 2 – went through a process of weighing the pros and cons and 3 – Took time to think things through. These actions words offer insight into how a person processes a scenario and thinks.

#9 Look for personality clues

Each and every one of us human beings has a unique personality, and these rudimentary classifications can enable us to assess and relate to another person. It also helps us read someone accurately. In looking for clues, you can ask:

- Did this person exhibit more introverted or extroverted behavior?
- Do they seem driven by significance or relationships?
- How do they handle risks and uncertainty?
- What drives them or feeds their ego?
- What kinds of mannerisms does this person exhibit when they are stressed?
- What are the kinds of mannerisms shown when they are relaxed?

By observing a person long enough, you can be able to pinpoint their base behaviors and mannerisms and set apart the odd one out.

CONCLUSION

It takes time to read people accurately. No one can identify a person's thoughts, feelings, and emotions just by looking at them and the singular mannerism and behaviors they exhibit. As we know, it could be something or nothing at all and it also depends on the context of the situation. Of course, there are exceptions to the rule but by keeping the basics of analyzing people and the principles with which you have built your powers of observation, you will eventually have greater control in reading other people, communicating with people effectively and understanding their thinking.

CHAPTER 8

HOW TO SPOT INSECURITY

WHEN SOMEONE IS BEHAVING IRRATIONALLY, you have to remind yourself that this could be because they are acting out of a certain emotion, or it also could be that their insecurity is behind this false sense of bravado. When you notice this, you will more likely procure a sense of empathy for these people who act arrogantly or rudely due to the fact that what they are trying to do is covering their insecurity.

Their insecurity can be about anything—looks, power, money, smartness, getting better grades, and so on—and most of these insecurities creep out from a sense of material value. Sometimes, insecurity can be justified—but most of the time, it is not. Insecurity manifests differently, and it can range from the inability to accept that they've done a great job or accept a compliment to as far as not wanting to wear a swimsuit to the beach.

FACTORS DETERMINING GOOD AND BAD

None of these traits helps us to behave virtuously. There is a fine line between being insecure and being a brat. Here are some identifying factors that can help you separate the good and the bad:

1. Self-kindness is not self-judgment.

Compassion towards someone who is insecure is being understanding and warm to them when they fail, or when we suffer or at moments when we feel inadequate. We should not be ignoring these emotions or criticizing. People who have compassion understand that being human comes with its own imperfections and failing is part of the human experience. It is inevitable that there will be no failure when we attempt something because failure is part of learning and progress. We will look into how failure is a friend in disguise in the next chapters. Having compassion is also being gentle with yourself when faced with painful experiences rather than getting angry at everything and anything that falls short of your goals and ideals.

Things cannot be exactly the way it should be or supposed to be or how we dream it to be. There will be changes and when we accept this with kindness and sympathy and understanding, we experience greater emotional equanimity.

2. Common humanity and not isolation

It is a common human emotion to feel frustrated especially when things do not go the way we envision them to be. When this happens, frustration is usually accompanied by irrational isolation, making us feel and think that we are the only person on earth going through this or making dumb mistakes like this. News flash— all humans suffer, all of us go through different kinds of suffering at varying degrees. Compassion involves recognizing that we all suffer and all of us have personal inadequacies. It does not happen to 'Me' or 'I' alone.

3. Mindfulness is not over-identification.

Compassion needs us to be balanced with our approach so that our negative emotions are neither exaggerated or suppressed. This balancing act comes out from the process of relating our personal experiences with that of the suffering of others. This puts the situation we are going through into a larger perspective.

We need to keep mindful awareness so that we can observe our own negative thoughts and emotions with clarity and openness. Having a mindful approach is non-judgmental and it is a state of mindful reception that enables us to observe our feelings and thoughts without denying them or suppressing them. There is no way that we can ignore our pain and feel compassion at the same time. By having mindfulness, we also prevent over-identification of our thoughts and feelings.

DISCOVERING COMPASSION

You're so dumb! You don't belong here loser! Those jeans make you look like a fat cow! You can't sit with us! It's safe to say we've all heard some kind rude, unwanted comments either directly or indirectly aimed at us. Would you talk like this to a friend? Again, the answer is a big NO.

Believe it or not, it is a lot easier and natural for us to be kind and nice to people than to be mean and rude to them whether it is a stranger or someone we care about in our lives. When someone we care is hurt or is going through a rough time, we console them and say it is ok to fail. We support them when they feel bad about themselves and we comfort them to make them feel better or just to give a shoulder to cry on.

We are all good at being understanding and compassionate and kind to others. How often do we offer this same kindness and compassion to ourselves? Research on self-compassion shows that those who are compassionate are less likely to be anxious, depressed or stressed and more resilient, happy and optimistic. In other words, they have better mental health.

IDENTIFYING SOMEONE WITH INSECURITY

Based on what was discussed above, when we are able to identify when a person is acting out of insecurity can enable us to protect ourselves from engaging in a mindless power play and feel insecure ourselves. People who are insecure tend to spread their negativity and self-doubt to others as well and here is how you can identify them and decide whether to show compassion or to show them the exit:

#1 People who are insecure try to make you feel insecure yourself.

You start questioning your own ability and self-worth and this happens when you are around a specific person. This individual has the ability to manipulate you and talk about their strengths and how they are good in this and that and in a way try to put you down. They project their insecurities on you.

#2 Insecure people need to showcase his or her accomplishments.

Inferiority is at the very core of their behavior and for people like this, compassion to tell them that they are not what they think in their heads is just a waste of your time. They feel insecure and to hide it, talk about their accomplishments, not in a good way but constantly brag about their amazing lifestyle, their wonderful shoes, their huge cars, and their elite education. All of this is done to convince themselves that they really do have it all and you have none.

#3 People who are insecure drops the "humble brag" far too much.

The humblebrag is essentially a brag that is disguised as a self-derogatory statement. In this social media age, you can see plenty of humblebrags who complain about their first-world problems such as all the travel they need to do or the amount of time they spend watching their kids play and win games or even the person who complains about having a tiny pimple when the rest of their face looks flawless. Social media is ripe with people who are narcissistic and this is not worth your time. Do not feel any less just because someone shows off how much of traveling they need to do.

#4 People who are insecure frequently complain that things aren't good enough.

They like showing off the high standards that they have and while you may label them as snobs, it might be a harder feeling to shake off because you might be thinking that they are really better than you although you know that it is all an act. They proclaim their high standards to assert that they are doing better than everyone else and make you feel less of yourself and more miserable. Pay no attention to people like this.

CONCLUSION

It does make sense that people who have better self-esteem and compassion as if you are happier and optimistic about your own future without having to worry about what insecure people have to say. When we continuously criticize ourselves and berate ourselves because we think other people are winning at life, we end up feeling incompetent, worthless and insecure ourselves which is what these people want us to feel. This cycle of negativity is vicious and will continue to self-sabotage us and sometimes, we end up self-harming ourselves.

BUT WHEN OUR POSITIVE INNER VOICE TRIUMPHS AND PLAYS the role of the supportive friend, we create a sense of safety and we accept ourselves enough to see a better and clear vision. We then work towards making the required changes for us to be healthier and happier. But if we do not do this, we are working ourselves towards a downward spiral or chaos, unhappiness, and stress.

CHAPTER 9

HOW TO SPOT ROMANTIC INTEREST

IF WE HAD the definite guide to spot a romantic interest, Tinder would go broke. That said, it is not hard to identify the telltale signs if someone is interested in you. Granted that some people are oblivious to it—but if you really do focus, you'd come to the realization if that person is indeed romantically interested in you or if they are just being flirtatious.

USUALLY, THAT SPECIAL SOMEONE STARTS WITH A CASUAL acquaintance, which leads to friendship—and before you know it, you look at this friend in a different light and keep thinking about them. Do they feel the same way you feel? Identifying if someone is interested in you romantically requires the careful and skillful interpretation of signals and actions.

WAYS TO FIGURE OUT IF SOMEONE IS ROMANTICALLY INTERESTED

Here are 15 ways to figure out if someone is romantically interested in you or if they are just flirting for the thrill of it:

#1 THEIR CONVERSATIONS WITH YOU

CONVERSATIONS, MEANINGFUL ONES ARE ONE OF THE WAYS A person shows a deeper interest in you and what you do. Do they keep asking you questions in an attempt to keep the conversation going? Pay attention to the questions they ask because it can tell you if they are genuinely showing interest in the things you do and like. A good and long conversation about your likes, dislikes, favorite music and so on is a classic sign of someone genuinely liking you and your company. If you are enjoying the conversation and the other person is engaging in it without looking bored or yawning, this is a sign that both parties are equally interested in each other.

#2 THEY KEEP BUMPING INTO YOU.

CALL IT FATE BUT THIS CAN ALSO BE A SIGN THAT THEY LIKE you and they are engineering any possible opportunities to meet you. This is sweet but also can be creepy if it becomes too much like stalking. If you feel that this person is following you or you suddenly feel uncomfortable, listen to your gut feeling and make a report. Stalking is serious and dangerous. However, if it bumping into you happens to be at places like the cafeteria or the lunch-room or neighborhood coffee place and not specific places like your gym that you've been going to for years, your house or anyway specific and private – make a complaint.

#3 They discuss future plans.

Another sign that someone could be romantically interested in you is if they plan for more dates or start talking about the near future because they clearly see you in it. It isn't about plans of getting married or buying a house but merely simple things like a concert in your area that they'd like to take you or even a friend's party in a week's time that they'd like you to come with. They have these upcoming events and they'd like you to be part of it.

#4 Five more minutes

If someone is interested in you, chances are they would like to spend a few more minutes longer with you. They don't mind adjusting their schedule just so they can spend an extra 5 more minutes to talk to you or even spend that extra 5 minutes on the phone just so they can continue talking to you. The fact that they do this is also an indication that they have romantic feelings for you.

#5 Reasons to spend time together

'I'm in the area—want to grab a bite?' or 'Oh you're having a cold? I can make a mean chicken soup—I'll bring it over' or even 'What are you doing right now? Want to go have dinner together?' Make no mistake that these could just be that the person likes spending time with you simply because you are a cool person to hang out with but if these reasons keep piling up and it only involves just the two of you, it is probably a big sign that this person likes you.

#6 Observe their body language.

If someone likes you, they mirror your body language and your movements. They sit in closer, they lean in, they smile when you smile, they find ways to touch you (not in a creepy way) like brushing against your shoulder, putting a strand of your hair behind your ear – all these are classic flirtation signs and if you are uncomfortable, say so but if you are enjoying it, this person is clearly into you.

#7 The compliments are mountainous.

Complimenting someone excessively can be a sign of ass-kissing or just trying to be nice. But if this person compliments you sincerely, it could be that they are interested in you. Look out for verbal cues such as complimenting your fashion choice or the way you style your hair. It could be that they are just being friendly, but them dropping compliments every time you meet is a big sign of them being interested in you.

#8 They remember the little things.

THE CLOSER YOU GET TO KNOW SOMEONE—THE MORE information you divulge to them. Your romantic interest will pick up a lot of interesting things about you and save it in their long-term memory and these things can be your favorite color, your favorite ice cream flavor, the first movie you watched together, where you first met – all of this is an indication that this person is genuinely interested in you.

#9 CONVERSATION STARTERS

SOME PEOPLE ARE SHY AND ARE NOT BIG TALKERS SO WHILE this is something to take note of, you cannot be the only one initiating contact all the time. If someone is willing to connect despite them being shy, that means they really do want to talk to you. Having one-way initiations for everything is a definite NO that the other person doesn't like you and do not see the need to spend the time to talk or even meet you but if they initiate contact as much as you do, that is a sure sign that they are into you.

#10 OTHER PEOPLE ARE OFF-LIMITS.

TAKE NOTE OF WHEN A PERSON TALKS ABOUT SOMEONE ELSE— do they talk a lot about other girls or guys when they are with you? Or is the conversation focused on just you and your person? What a person says in a conversation and how they refer to other people in their social circle can give you real clues into whether they are romantically interested in you. Talking about going on a date with a girl or guy is not really a good indication that this person likes you.

TRUSTING YOUR FEELINGS AND YOUR INTUITIONS IN ALL these possible scenarios above is the best bet. Remember that different people do different things to show someone they care or that they are interested in them and cultural values, upbringing, and societal norms also play a big part in identifying these signs so nothing is set in stone. All the signs described above are a good telling sign that a person is interested in you especially if they like spending more and more time with you. Even if you are not sure, you can exhibit signs that you are interested in them so that they will also have an idea but to be on the safest side, telling someone that you like them and you'd like to get to know them better and even start dating is the best way forward to prevent any miscommunication or misunderstanding between two people.

OF COURSE, THE GAME OF LOVE IS NOT AS STRAIGHTFORWARD and as simple as it is. It takes a little bit of dating experience to figure out if someone is into you or not or you can just do the good old fashion trial and error, get your heart broken, kiss all the toads till you meet your prince or princess charming.

CHAPTER 10

HOW TO SPOT A LIE

BODY LANGUAGE IS an important part of communication, and both visual and verbal elements of it should align with the message you are sending out—otherwise, people will find it easy to see that you are being deceptive. There are definitely clues in behavior that you will be able to identify if a person is telling a lie or even just shielding the truth. Nobody likes a liar, and nobody likes being lied to. It is heartbreaking to know that someone is lying to you, especially if it is a partner. If it is in a professional setting, all trust is gone out the window with even one tiny lie. Lying and the extent of it makes and definitely breaks relationships—depending on what the lie is.

SCENARIOS OF LYING

As much as we want to say we always tell the truth, the realities of human communication are complex. Here are some scenarios of lying:

1 – Some lies are intended as a courtesy or are habitual.

For example, when you say 'I'm fine' when someone asks how you are doing and while you may not be fine, you just say it anyway to prevent them from asking you more questions or to be involved in more conversation. Most adults lie plenty of times, every single day and while this is fine, some lies though will get you into trouble.

2 – Some forms or lies are even expected.

It has become so habitual to lie in our culture that we become to expect it. Legal strategies used is a scenario of when 'plausible deniability' is expected and customary. We always face scenarios such as cross-company relationships and also adherence to nondis-closure agreements. Certain situations require fast thinking and practice to keep to your commitments while staying honest about the information we want to protect. So when asked 'What is your weakness?' you tell a fib that makes a negative trait a positive one.

3 – Lying by deflecting

On a regular basis, we are bombarded with people who by nature or through training avoid showing off excessive body language and deflect core questions and thus, flood their audiences with irrelevant information or use deceptive forms of truth. Take politicians for example. So while a deception can be easily spotted in action, it is much more difficult to spot a deception with a person who is fundamentally dishonest.

KEY ELEMENTS IN SPOTTING A LIAR

So while there are these various forms of lying that is done on a daily basis, what if there is a person that you really want to know if they are lying or not? According to Susan Carnicero, a CIA officer, one can usually tell if a person is lying in 5 seconds. So if you want to analyze someone or read someone, here are some key elements that you need to pay exceptional attention to know how to spot a liar.

#1 Analyzing versus speculating

As we talked about in chapter 1 about how when a person crosses their arms, it could mean that they are uncomfortable or it just could mean that they are cold or it's just a habit of theirs. Assuming that crossed or folded arms is a signal of deceptive behavior is just speculation. You should instead analyze whether this behavior or mannerism is a result of the question asked. Knowing the first sign of deceptive behavior will happen in the first five seconds of the question asked will enable you to determine if that question was the one that produced the folded arms. This first clue of deception could even happen while the first question is being asked, which goes to know that this person's brain is moving much faster than the words coming out of the interviewer – it is a sign that they are subconsciously trying to frame their response. You should also look for clusters of behavior and whether these clusters are a direct response to the question and not just a nervousness bout.

#2 Managing your bias

Believe it or not, people who are being deceptive can give you truthful answers. They give you true information thus increasing your belief that they are telling the truth and simultaneously, lessen your ability to identify when they are in fact lying. To prevent this, you need to focus on filtering out the fluffy, truthful responses they give. So really focus on what the fluff in so that you won't be caught by surprise thinking that this person seemed honest the entire time.

#3 Recognizing when someone is being evasive

When a person is being deceptive, they usually create fluff and give long explanations of their answers without actually addressing the issue. They are adept at deflecting and redirecting their responses such as 'Have I ever been accused of doing this before?', 'Do have a reputation for doing good?'. When someone goes on and one for a good 10 minutes and never answers the question this is a sign that they are lying.

#4 Watching out for signs to deny

One of the most important things you need to listen to is the direct denial of an accusation. When a person is guilty, they will verbally create fluff, not getting to the point, not answering the question and attempt to justify the situation by saying things like 'not likely' or 'not for the most part' or the common favorite 'not really'. A person who responds to the question of 'Did you do this?' with 'This is not really the way we do things around here' instead of a definite 'NO, I did not', that is a big sign of deception.

#5 There will be signs of aggression.

Well when someone gets angry at the question being asked, it is 100% an indication that they are guilty. They may exhibit signs of aggression such as attacking the interviewer or they would also attempt to flip the situation by accusing the questioner of bias or discrimination. "Don't you see what he has done? Don't focus on just me!' or 'It's your fault for not keeping it this way, not mine' may be some ways of deflecting from answering the question. They exhibit anger, disgust or even go on to the blame game in response to the question asked.

#6 Convey versus convince

You ask a question and instead of a yes or no answer, the person launches into an amalgamation of how they are a good person, they have been a good employee, how long they've worked and have had no issues, how they have a good reputation – This person is providing an unsolicited statement to defend themselves instead of just answering the very direct question with a very direction answer. Disqualifiers added to their statements such as 'You know I won't lie to you' or 'You trust me, don't you?' or 'to be perfectly honest with you' or the crowd favorite 'Quite frankly are all attempts to prevent themselves from answering. They are subconsciously trying to cover up their lie and may also use race or religion to justify such as saying 'I swear on the bible' to create a compelling and convincing argument.

#7 Look out for nonverbal signals.

If you see an inappropriate amount of concern for a situation or the lack thereof or even a smile or a smirk in response to a question like 'Did you kill that man?', these are all nonverbal clues you need to pay attention to because they can be extremely subtle but very useful in catching a liar. These nonverbal clues extend to include jumping in to answer a question fast or even a pause. They also have a habit of not being able to look at the questioner in the eye or in contrast, stare with aggression to the questioner. They may even say 'No' but be nodding a 'Yes'. A person can even touch their face or nose or even cover their mouth or face because this is another subconscious way of hiding a lie. The stress of deception can also cause the skin to turn cold and start itching or even flush – notice when they suddenly scratch their ears or nose. Look out for anchor point movements such as the changes in the arms or even the feet. A foot could be dangling but suddenly tapping nervously or even pointing at a different direction.

Summary

All of these situations are important to watch, and you must also watch the cluster of behaviors and activity as opposed to zoning in on only one behavior. Spotting whether someone is telling a lie or the truth can be hard at first, and it requires training to efficiently tell if someone is, in fact, lying. Wrongly accusing someone can be disastrous not only to them but also to your reputation—so depending on the situation, look out for the body language, understand your baseline, and properly look for nonverbal clues.

CONCLUSION

At the end of this book, you are now better at analyzing and reading cues as well as become more adept at understanding yourself and the people around you. Knowing how to analyze people effectively is crucial in the business world as well as the social world, the political world, and the socio-political world. In fact, knowing how to do this the right way helps in just about any aspect of life.

If you're in the marketing department or a designer or presenting a project proposal or even meeting a patient—it is extremely important and valuable for you to recognize signals and cues of the people you meet. Through reading this book, it will help you get more accustomed and comfortable in analyzing people—however, never stop practicing to look for nonverbal and verbal cues, as this is what helps give you a lead.

Not only will this give you a better headway in life, but it will also help you create more meaningful and long-lasting relationships. It makes you a better friend, a better partner, a better co-worker, or even a better boss just by knowing what cues and gestures people use to convey their innermost thoughts—especially when words fail them.

By learning what non-verbal gestures mean, you would be able to break the code that would lead you into learning more about the people around you and empathize with them. At the same time, hopefully, this book has enabled you to improve your relationship with your friends, colleagues, partner, and family by learning how to communicate better through the right non-verbal gestures.

CPSIA information can be obtained
at www.ICGtesting.com
Printed in the USA
LVHW021327080920
665327LV00006B/1010